NASA'S SUPERSONIC REBIRTH

Transforming the Future of Flight with Echoes of Concorde, Breaking the Sound Barrier at Mach 2, and Unveiling the BOOM of Innovation

JOHN MARTINEZ

TABLE OF CONTENTS

INTRODUCTION

With the death of the Concorde in 2003, it is, of course, no longer possible to travel quickly across the Atlantic. It takes around eight hours to travel from London to New York and seven hours to travel from New York to London. With the help of a helpful jetstream, the current record time is little under five hours from New York to London.

But NASA has brought back the notion of supersonic flight, speculating that a trip from New York to London may take as little as ninety minutes in the future.

People are talking about this, as evidenced by the comments above.

However, the most important queries are: "What are supersonic planes? After the Concorde, why did supersonic aircraft manufacturing cease, and is it possible to rebuild them?

CHAPTER ONE:

MYSTERIES IN SWIFT, ROARING AIRCRAFT: WHAT ARE SUPERSONIC PLANES?

The American Bell X-1 experimental aircraft, which was propelled by a 6,000-pound (2,700 kg) thrust rocket driven by liquid oxygen and ethyl alcohol, was the first aircraft to fly supersonic in level flight. Supersonic aircraft have mostly been military or experimental in the past. There have been several claims of breaking the sound barrier resulting from the advent of rocket and jet-powered aircraft during World War II. On October 14, 1947, however, human aircraft achieved the first officially acknowledged controlled level flight that exceeded the speed of sound. In this historic accomplishment, Chuck Yeager piloted the experimental Bell X-1 research rocket plane. The F-86 Canadair Sabre, piloted by the first female supersonic pilot,

Jacqueline Cochran, was the first aircraft in production to breach the sound barrier.

According to David Masters, the DFS 346 prototype, which the Soviets captured in Germany and broke free from a B-29 at 32,800 feet (10,000 meters) height in late 1951, had a speed of 683 mph (1,100 km/h), which may have surpassed Mach 1 at that altitude. Wolfgang Ziese flew these flights as the pilot. On August 21, 1961, during a test flight at Edwards Air Force Base, a Douglas DC-8-43 (registration N9604Z) purposefully exceeded Mach 1 in a controlled descent. This was another noteworthy incident. The crew comprised of William Magruder (pilot), Paul Patten (copilot), Joseph Tomich (flight engineer), and Richard H. Edwards (flight test engineer). This was the first and is still the only intentional supersonic flight by a civilian aircraft outside of the Concorde or Tu-144.

During the 1960s and 1970s, various design studies for supersonic airliners were done, leading to the eventual introduction of two types into service: the Soviet Tupolev Tu-144 in 1968 and the Anglo-French Concorde in 1969. But a tragic Concorde crash, as well as political, environmental, and financial barriers, kept them from being utilized to their full commercial potential. One of the four flight speeds is supersonic. We refer to them as flying regimes. There are flying regimes that are subsonic, transonic, supersonic, and hypersonic.

Vehicles that travel at supersonic speeds surpass the speed of sound. The sound travels at about 768 miles per hour (1,236 kilometers per hour) at sea level. These speeds are expressed in terms of mach numbers. The aircraft's speed in relation to the speed of sound is known as its Mach number.

Any flight faster than Mach 1 is considered supersonic. Supersonic, often known as Mach 5, refers to traveling five times the speed of sound. The behavior of the air around an airplane as it travels through the atmosphere is determined by the aircraft's speed in relation to the speed of sound. In honor of physicist Ernst Mach, this ratio is referred to as the Mach number. A plane is said to be supersonic when it exceeds the speed of sound. Supersonic aircraft typically travel between 750 to 1500 miles per hour with a Mach number of one or higher.

The first aircraft to test supersonic flight was the Bell X-1A, which made its inaugural flight in 1947. It proved that people were capable of traveling faster than sound. Modern supersonic fighter planes are still influenced by the aerodynamics of these early supersonic aircraft. There have been unsuccessful attempts to build supersonic airliners that are affordable, such as the Russian TU-144 and the Concorde.

Because supersonic flight creates a lot of drag, high-thrust gas turbine engines power supersonic fighters like the F-14. The F-14's swept wings aid in reducing drag, and the pilot can modify the wing sweep for best results at various speeds. The airframe's frictional heating is sufficiently low for the use of lightweight aluminum for the structure at speeds lower than Mach 2.5.

What zooms faster than sound?

One example of a supersonic object is a gun bullet. Military fighter planes can also fly at this speed. The space shuttle orbiter flies at supersonic speeds for a portion of its voyage.

The most well-known passenger plane to reach supersonic speed was the Concorde. The Concorde reached a maximum speed that was over double the speed of sound. In just three and a half hours, it could transport passengers from London to New York. That is about half the time required for the same distance flown by a normal airplane. The Concorde is not currently in service. In 2003, it was shut down.

CHAPTER TWO

CONCORDE'S EPIC FLIGHT THROUGH HISTORY

The famous Franco-British supersonic airliner Concorde was developed as a result of research and development conducted jointly by the British Aircraft Corporation (BAC) and Sud Aviation (later Aérospatiale) starting in 1954. On November 29, 1962, the official development treaty was signed. It was projected that the program would cost £70 million (about £1.39 billion in 2021). The first flight of the six prototypes occurred at Toulouse on March 2, 1969, with construction starting in February 1965.

The ogival delta wing, drooping nose for better landing visibility, and narrow fuselage, which could accommodate 4-abreast seating for 92–128 passengers, were all features of the Concorde, which was built without a tail. Four Rolls-Royce/Snecma

Olympus 593 turbojets, outfitted with variable engine intake ramps and reheat for takeoff and supersonic acceleration, provided the vehicle with its power. Interestingly, the Concorde was the first aircraft to use analog fly-by-wire flying controls and was capable of maintaining a supercruise at 60,000 feet and Mach 2.04.

Concorde had difficulties, including delays and cost overruns that increased the program cost to £1.5–2.1 billion in 1976 (£9–13.2 billion in 2021), despite early optimism and a projected market for 350 aircraft. Nevertheless, Air France and British Airways launched it into commercial operation on January 21, 1976. Its backbone became transatlantic flights, especially to Washington Dulles and New York JFK.

Although Concorde greatly shortened transatlantic flight durations, its employment was limited to transoceanic routes due to sonic booms. The Boeing 2707, a possible competitor, was canceled in 1971, and its only adversary, the Tupolev Tu-144, had difficulties, including a crash in 1978.

On July 25, 2000, Air France Flight 4590 had a catastrophic accident shortly after takeoff, killing 113 people. This was the sole fatal Concorde event, and as a result, commercial operations were halted until November 2001. The 27-year commercial operation of Concorde came to an end in 2003 when it was retired, despite attempts to address safety concerns. Numerous Concorde aircraft are currently on exhibit at museums in North America and Europe, acting as enduring reminders of the advancements in aviation and supersonic flight.

A pact signed on November 29, 1962, formalized Britain and France's unprecedented partnership on the Concorde. This partnership featured British Aerospace and Aérospatiale handling the airframe, while Rolls-Royce and SNECMA produced the jet engines. As a result, on March 2, 1969, the stunning delta-wing Concorde made its first flight.

Flying from London to New York took only three hours thanks to the Concorde's maximum cruising speed of Mach 2.04, or 1,354 miles per hour. The Concorde's enormous development expenses prevented it from ever turning a profit, despite being a technological marvel. But it also demonstrated how well European manufacturers and governments collaborated on challenging projects, highlighting Europe's supremacy in aerospace development.

The Concorde's attributes

With over 50,000 flights completed at supersonic speeds, the British Airways Concorde carried over 2.5 million passengers. It could travel from London to New York in less than three and a half hours at a top speed of 220 knots (250 mph) and cruise at 1350 mph, which is far quicker than subsonic trips, which required around eight hours. A British Airways Concorde made a full round of the earth in November 1986, traversing 28,238 miles in less than 30 hours.

Using'reheat' technology, the Concorde, which had the strongest pure jet engines in commercial aircraft, added fuel to increase power during takeoff and the transition to supersonic flight. On February 7, 1996, it accomplished the quickest transatlantic journey in 2 hours, 52 minutes, and 59 seconds, traveling from New York to London.

Due to airframe heating, the Concorde's length, which was about 204 feet, expanded between 6 and 10 inches while in flight. These alterations were managed and the heat produced by supersonic flight was dispersed with the use of specially created white paint. Before being certified for passenger flights, the Concorde underwent 5,000 hours of testing by a committed team of over 250 British Airways engineers working closely with regulators. This made the aircraft the most tested in history.

Some of its characteristics are as follows

<u>Engine</u>

Four Rolls-Royce/Snecma Olympus 593 Mk610 turbojets underneath the Concorde's striking delta wings provided the aircraft with its incredible supersonic prowess. The RAF's Avro Vulcan strategic bombers were powered by Rolls-Royce Olympus engines, which served as the model for these engines.

The Olympus 593 engines of Concorde were no ordinary miracles, as they drew influence from the Vulcan's high-altitude achievements and delta wing design. They began life as the Bristol B.E. 10 and became the first two-spool axial-flow turbojets in history, making their debut in May 1950. The Concorde's afterburners, which increased thrust during takeoff and supersonic flight, added to the aircraft's technical superiority.

When the Concorde was operating "dry," or without afterburners, each of its four engines produced

31,000 pounds of thrust. Using the afterburners, or operating in a "wet" state, increased this output by more than 20%, resulting in an astounding 38,050 pounds of thrust per engine.

The Concorde's engine technology was revolutionary, even though its maximum takeoff weight was just 185 tonnes, far less than the 333 tonnes of the Boeing 747-100. With this ability, the elegant aircraft was able to "supercruise" at speeds more than twice the speed of sound, generally flying at around 2,158 km/h (1,165 knots), which was only slightly faster than its top Mach 2.04 speed.

<u>A pair of wings</u>

The Concorde's look was a visual spectacle that distinguished it from modern aircraft as well as its subsonic predecessors. Unlike the straight-edged designs of other fighter planes, its unique wings, called an ogival delta, were distinguished by a curved leading edge.

Because Concorde skillfully used the delta wing's benefits in high-altitude supersonic flight, the aircraft type has become quite popular in military aviation. The narrow wings of the Concorde reduced drag and improved its aerodynamics in contrast to the wider wings of its competitors.

The Concorde created shockwaves when it broke through the sky at supersonic speeds, but oddly, this was to its benefit. High pressure was produced under the wings by these shockwaves, which increased lift without increasing drag. This relationship was important for both altitude and speed. Concorde reached remarkable altitudes, maximizing its supersonic efficiency by taking advantage of the lower air resistance in the thinner environment.

<u>Nose</u>

The movable, drooping nose of the Concorde was a unique feature that greatly contributed to the aircraft's optimal performance during both cruise and landing. A smooth, streamlined front profile with less surface area was produced when the nose was positioned directly from the cockpit, which decreased drag and allowed for faster speeds.

Nevertheless, Concorde had a very steep angle of attack throughout the landing phase. Pilots would have very poor sight if the nose was pointing straight forward at this crucial point. The worries about takeoff and taxiing were the same. The Concorde's nose could be lowered by an amazing 12.5° to improve vision before to landing in order to remedy issue. This angle was reduced to 5° during landing to reduce the possibility of damage occurring when the nose wheel made contact with the runway. This clever nose design demonstrated how adaptable the Concorde was, providing maximum performance on the ground as well as in the air.

<u>Paint</u>

The paint finish of the Concorde, for example, is one of those apparently little touches that significantly improved the plane's performance. In particular,

Concorde's white paint was purposefully very reflective. This was a very wise decision since it assisted in deflecting some of the heat produced during supersonic flight.

More than simply a cosmetic issue, being able to withstand this heat was essential for avoiding overheating and protecting the aluminum backbone of the airplane. Concorde was able to safely sustain supersonic speeds for lengthy periods of time, guaranteeing structural integrity and safety, because to its reflecting coating. For instance, a Concorde with the Pepsi logo and a promotional blue livery had a 20-minute maximum flight time at supersonic speeds because of its less heat-resistant paint.

CONCORDE'S LIMITATIONS: SKY'S SUPERSONIC CHALLENGES

Twenty years after Concorde's remarkable 25-year career came to an end, the aircraft still piques the interest of a generation that was never able to see its miracle. This interest endures even among tech-savvy millennials already inundated with cutting-edge inventions in a century marked by caution, environmental concern, decarbonization, and economic viability. Why does this legendary airplane still live on in people's minds?

Let's address the complaints that are often directed on Concorde, trying to write it off as a relic of the past with derogatory words. Yes, it was noisy; yes, it used gasoline at a rate that only the wealthy could afford; yes, it presented problems for operational profitability; and yes, there was a regrettable accident. But each of these criticisms merits more investigation; are there any subtleties that provide a more impartial viewpoint?

Admirers were drawn to windows and doorways by its perfect silhouette and thundering, magnificent roar, which announced its departure or return. Although the decibel levels may make modern ears twitch, the deliberate application of the loud noise was courteous and limited to boats and birds.

Because of the incredible speed and shock waves required for supersonic travel, there was a noticeable increase in energy consumption, which led to the burning of more kerosene—a fact that was required by the indisputable rules of physics. This did add to the contamination of the atmosphere by bringing in nitrogen compounds, carbon monoxide, and ozone. However, Concorde was not the only offender among the many subsonic planes that fly all over the world or the many rockets that are fired every year into clouds of poisonous smoke.

This obviously had a price, making boarding available only to the wealthy. But the common 'wealthy' term is oversimplified. In addition to leisure travel, Concorde faced competition from business aviation and the upscale cabins of subsonic aircraft, which catered to time-pressed professionals. In an age of fast travel, the Concorde was an efficient vessel.

Concorde's old-fashioned equipment and just 14 aircraft in operation made it distinct from the frequently upgraded subsonic jets. The transition to digital technology also resulted in the cost and obsolescence of Concorde spare components.

The significance of Concorde was then eclipsed by a terrible catastrophe. Aviation regulations changed as a result, although it's important to remember that Concorde had the correct certification at the time.

Supernatural Expedition

Throughout its development and operation, the Concorde, a supersonic passenger jet powered by turbojets, was envisioned as the aircraft of the future.

The project was started in the 1950s, and in 1969—the year of the Apollo 11 moon landing—it made its first flight. A public demonstration followed in the 1970s. Up to its closure in 2003, the operation remained in place.

One of Concorde's benefits was that it took less time to fly. Heathrow to New York, Virginia, and Barbados flights, for instance, might take less than half as long as those on other airlines.

That being said, only the very rich could afford the $7,995 round-trip ticket price from New York to London in 1997. In addition to the outrageous fees that customers paid, the program's operating costs for the supersonic transportation came to an astounding £1.3 billion.
A number of restrictions were put in place due to the amount of noise that Concorde flights made.

Due to these issues, the aircraft was no longer profitable, and British Airways and Air France jointly announced its retirement in 2003.

Travel Cost

As the Concorde soared through technical advancements, providing a luxury flying experience, low-cost airlines were emerging as another aviation trend.

When there were upscale services like Concorde, flying was considered a luxurious and expensive experience in the past. 'No-frills' flights were popular at the same time.

Not only the rich elite could now afford to travel by air, thanks to developments that began in the 1960s. Consumers were prepared to give up certain benefits in exchange for less expensive travel; the final destination was more important than the route.

Over the last 40–50 years, this tendency hasn't really altered all that much. The duration of flights is still about the same, and budget travel is still as popular as it was in the 1960s and 1970s.

But aerospace technology is always evolving, with an emphasis on environmentally friendly planes.

Here in the UK, we're experiencing these developments directly.

With a mixture of difficulties and achievements now, there's a chance that people may forget about Concorde. Even the most incredible aircraft become forgotten in the vast realm of aviation. Despite its significance, the Concorde's impressive advancements might be overshadowed by Airbus's greater success, making it simple to forget about this revolutionary period in aviation history. However, it seems like NASA is going to refute that! Maybe supersonic flight will make a reappearance in the future, or maybe something quite different.

CHAPTER THREE

IS NASA PLANNING TO BRING BACK SUPERSONIC TRAVEL?

Imagine what's making a return on the aviation scene? NASA is spearheading the development of supersonic travel, with the goal of reducing travel time from New York to London to as little as ninety minutes.

NASA recently revealed their investigation of commercial planes that are traveling at an incredible Mach 4, or more than 3,000 miles per hour, in a blog post on the "high-speed strategy." Between 50 and 60 possible routes, most of them transoccanic (passing across the North Atlantic and Pacific), were found in their Glenn Research Center analysis. Banned in countries like the US, supersonic overland travel is presently prohibited.

What's really interesting is that NASA is developing the X-59, a "quiet" supersonic aircraft, as part of the Quest mission. the objective? to maybe persuade authorities to reevaluate the regulations and allow flights between Mach 2 and Mach 4 (1,535 - 3,045 miles per hour). Let's put the former supersonic champion Concorde in perspective. Its maximum speed was Mach 2.04, or 1,354 miles per hour. Picture yourself aboard a jet zipping at Mach 4, and ninety minutes after taking off from New York, you might be having tea in London.

NASA is accelerating its efforts to achieve high-speed flight! The Advanced Air Vehicles Program (AAV), having finished recent research, is preparing for the next stage. Together, they have been tasked with creating concepts that would enable Mach 2-plus travel in collaboration with Boeing and Northrop Grumman Aeronautics Systems. At supersonic speeds, picture yourself flying!

There have been previous NASA experiences in the supersonic domain. The groundbreaking X-59 aircraft was made possible ten years earlier by comparable investigations. These recent findings are now expected to update technological roadmaps and identify further high-speed range research requirements.

Thoughts of efficiency, safety, and social concerns are also taken into account in addition to speed. Hypersonic Technology Project manager Mary Jo Long-Davis stresses the need of responsible innovation. They are considering more than simply the next exciting thing.

The NASA X-59 test aircraft, which Lockheed Martin completed in July, is designed to reduce sonic booms to gentle thumps and perhaps pave the way for overland supersonic travel. In 2027, NASA intends to provide regulators with sufficient data, and ground testing and the first flight are to follow.

Supersonic Marvel: The X-59

In addition to breaking the sound barrier, NASA's experimental supersonic wonder, the X-59 Quesst, is doing it in style thanks to a striking red, white, and blue makeover. This aircraft, which was once green, is scheduled to make its inaugural flight where it will effortlessly break over the sound barrier. It now sports a sleek white fuselage with a sonic blue underbelly and dramatic red wing accents.

However, there's more to this patriotic paint job than meets the eye. It is functional in that it protects the X-59 from moisture and rust. It also has safety markers that are essential for operations both on the ground and in the air. Recent relocation to the paint barn at Lockheed Martin Skunk Works in Palmdale, California, was a major milestone in this mission of unprecedented altitude.

Project manager for the Low Boom flight demonstration Cathy Bahm said she was excited about this revolutionary stage, saying, "I expect the moment to take my breath away because I'll see our vision coming to life when the X-59 emerges from the paint barn with fresh paint and livery." Big things are expected of the X-59 in the next year, and its vivid exterior now corresponds with the amazing mission it is about to undertake.

The goal of the X-59's design, which includes both supersonic speed and reducing acoustic disruptions, is to breach the sound barrier with a soft "thump" rather than the usual loud boom. This revolutionary aircraft is being developed at Lockheed Martin's Skunk Works facility in Palmdale, California. When it is prepared, the X-59 will fly over a few American villages to collect vital information on the noise pollution that the locals face.

With its X-59 program, NASA is making great progress toward a contemporary supersonic aircraft that will show that conventional supersonic booms can be reduced to mere "thumps," opening the door to the prospect of commercial supersonic flights over land. An innovative eXternal Vision System (XVS) was put through a rigorous flight test in August of last year, and now it is being put through a high-frequency shaking test to make sure it can withstand the rigors of in-flight circumstances.

In non-supersonic aircraft, the XVS, which takes the place of a conventional windshield, has shown value in providing real-time visibility during test flights. However, NASA needs to guarantee it can resist the strains of flying faster than the speed of sound aboard the X-59 itself. In order to verify the system's robustness, continuous shaking testing subject it to vibrations like to those encountered during flight.

An essential step prior to the XVS being installed on the X-59 is this pre-qualification testing. NASA hopes to demonstrate how this technology might completely transform commercial aviation once temperature and altitude testing is finished.

If the X-59 program is successful, it may open the door to revising the regulations that now prohibit supersonic flights over land, bringing in a new age of environmentally friendly and quieter aviation.

CHAPTER FOUR

BRACING FOR SUPERSONIC FLIGHT WONDERS: ARE WE READY FOR SUPERSONIC PLANES

The competition for ground-breaking breakthroughs in aviation technology is intensifying, and it goes beyond the X-59. Hold on tight because, according to the newest buzz from Britain's Civil Aviation Authority (CAA), we may be flying from London to Sydney in only two hours by 2033. We're talking about suborbital "Earth to Earth" journeys, when rockets propel humans up to 125 miles into space at a whooping 3,500 mph. Forget about the present 22-hour marathon.

Imagine Virgin Galactic and Jeff Bezos' Blue Origin combined on steroids. With SpaceX's Starship rocket, which is expected to transport 100 people between continents in less than an hour by 2025, Elon Musk is already playing the game. Not only that, but Tianxing I, China's "rocket with wings," is preparing for suborbital testing with the goal of achieving a crewed flight by 2025, which may allow it to go an astounding 4,300 miles in an hour.

Big names in aviation are betting on suborbital space travel; UBS estimates that if 5% of long-haul flights choose this route, the industry could grow to an astounding $20 billion annually. The subsonic market has the potential to reach $805 billion by 2030. Pretty exciting, huh?

However, and there's always a but, suborbital missions encounter several difficulties. Safety comes first, particularly when it comes to rockets that run on fuels like liquid natural gas or the explosive combination of liquid oxygen and liquid methane. Although Elon Musk's Starship promises to move at breakneck speed, explosions are not part of the plan.

Not to mention the runway, which is the big problem in the room. These innovative aircraft are still in the testing stage and are overcoming lengthy regulatory obstacles before they ever take to the air. After all, it took the Concorde 20 years from engine testing until its first flight.

When it comes to obstacles, sustainable aviation fuels, or SAFs, are moving slowly forward. They may cost up to eight times more than conventional jet fuel and now make up a pitiful 0.1% of aviation fuel. The US government is aiming for 100% SAFs by 2050, but a new analysis highlights obstacles, including a lack of biofuel supply and raising concerns about the sustainability of certain feedstocks.

While the pursuit of greener sky is admirable, it's not easy. Consider Enviva, the biggest manufacturer of wood pellets in the world, which converts trees into biofuels. Sounds sustainable, doesn't it? The carbon held in the wood is released when that fuel is used, thus tree replanting may not keep up. It is, very literally, a net-zero claim in the air.

And then there's the audacious Prometheus, who claims he can use CO2 extracted from the atmosphere as fuel. It sounds idealistic, but critics claim it's untested and unreviewed, too good to be true. Supersonic aircraft have efficiency issues even with the greenest fuel; they may use up to nine times as much fuel per passenger per kilometer as subsonic flights, endangering the environment and their economics.

Ultimately, achieving the goal of supersonic travel may need a great deal of patience. NASA's initiatives and developing technology have the potential to transform subsonic aircraft, providing promise for a more efficient and clean future. Hopefully the tough ascent will be worth the view, but let's hope not.